DIG DEEP INTO FOSSILS

Digging Up PLANT FOSSILS

Enslow PUBLISHING

BY CHARLOTTE TAYLOR

INVESTIGATE!

Please visit our website, www.enslow.com. For a free color catalog of all our high-quality books, call toll free 1-800-398-2504 or fax 1-877-980-4454.

Library of Congress Cataloging-in-Publication Data

Names: Taylor, Charlotte, 1978- author.
Title: Digging up plant fossils / Charlotte Taylor.
Description: New York : Enslow Publishing, 2022. | Series: Dig deep into fossils | Includes index.
Identifiers: LCCN 2020030900 (print) | LCCN 2020030901 (ebook) | ISBN 9781978521681 (library binding) | ISBN 9781978521667 (paperback) | ISBN 9781978521674 (set) | ISBN 9781978521698 (ebook)
Subjects: LCSH: Plants, Fossil–Juvenile literature.
Classification: LCC QE906 .T39 2022 (print) | LCC QE906 (ebook) | DDC 561–dc23
LC record available at https://lccn.loc.gov/2020030900
LC ebook record available at https://lccn.loc.gov/2020030901

Published in 2022 by
Enslow Publishing
101 West 23rd Street, Suite #240
New York, NY 10011

Designer: Katelyn E. Reynolds
Editor: Megan Quick

Photo credits: Cover, pp. 1 (plant fossil) Artur Balytskyi/Shutterstock.com; cover, pp. 1–24 (background) Click Bestsellers/Shutterstock.com; cover, pp. 1–24 (tools) Serz_72/Shutterstock.com; cover, pp. 1–24 (burst) Foxys Graphic/Shutterstock.com; p. 4 Faba-Photograhpy/ Moment/Getty Images; p. 5 joakimbkk/ E+/Getty Images; pp. 5, 10, 11, 13, 15, 24 DeAgostini/Getty Images; p. 7 Nagendra Yadav/ iStock / Getty Images Plus; p. 8 Succu/Wikipedia.org; p. 9 Wilson44691/Wikipedia.org; p. 12 madsci/ iStock / Getty Images Plus; p. 14 aykut karahan/ iStock / Getty Images Plus; p. 16 Plantsurfer/Wikipedia.org; p. 17 Jonathan Blair/ Corbis NX / Getty Images Plus; p. 18 Lyn Alweis/The Denver Post via Getty Images; p. 19 CribbVisuals/ iStock / Getty Images Plus; p. 19 Terry c3D/ iStock / Getty Images Plus; p. 21 Alinabel/ iStock / Getty Images Plus; p. 22 Geography Photos/Universal Images Group via Getty Images; p. 23 EdeWolf/ iStock / Getty Images Plus; pp. 25, 26 Wild Horizons/Universal Images Group via Getty Images; p. 27 Demetrio Carrasco/Getty Images; p. 27 alice-photo/ iStock / Getty Images Plus; p. 27 Scientifica/ Corbis NX / Getty Images Plus; p. 28 SAUL LOEB/AFP via Getty Images; p. 29 arifoto UG/picture alliance via Getty Images.

Portions of this work were originally authored by Kathleen Connors and published as *Plant Fossils*. All new material in this edition authored by Charlotte Taylor.

Printed in the United States of America

Some of the images in this book illustrate individuals who are models. The depictions do not imply actual situations or events.

CPSIA compliance information: Batch #CSENS22: For further information contact Enslow Publishing, New York, New York, at 1-800-398-2504.

CONTENTS

WORDS IN THE GLOSSARY APPEAR IN **BOLD** TYPE THE FIRST TIME THEY ARE USED IN THE TEXT.

THE BEGINNING OF PLANTS

A plant can be many different things, like a flower, bush, or tree. Over 440 million years ago, the first plants appeared on land. Scientists believe that they **evolved** from green **algae**. It took a long time for the first plants to become the flowers, bushes, and trees that you know today.

ALGAE FLOATING IN WATER

Explore More!

ALGAE LIVE MOSTLY IN WATER. THE FIRST PLANTS PROBABLY LIVED IN WATER TOO.

How can we know what happened millions of years ago? Fossils! A fossil is the **remains** of something that died a long time ago. Scientists can study plant fossils to learn what plants used to be like and how they have changed over the years.

LIVING OAK LEAVES

A FOSSIL, LIKE THIS OAK LEAF, MAY BE HUNDREDS, THOUSANDS, OR MILLIONS OF YEARS OLD.

A PLANT BECOMES A FOSSIL

There are a few steps to making a plant fossil. First, the plant is buried in **sediment** when it dies. Plant fossils are often found near bodies of water where there's a lot of moving sediment, such as lakes and rivers. Then, over many years, the sediment turns to rock. This **preserves** the plant inside.

Plant fossils aren't as common as animal fossils. The plant fossils found most often include **pollen**, **spores**, and large, woody plants that don't break down easily. Scientists have found fruit and seed fossils too.

THE PARTS OF A PLANT

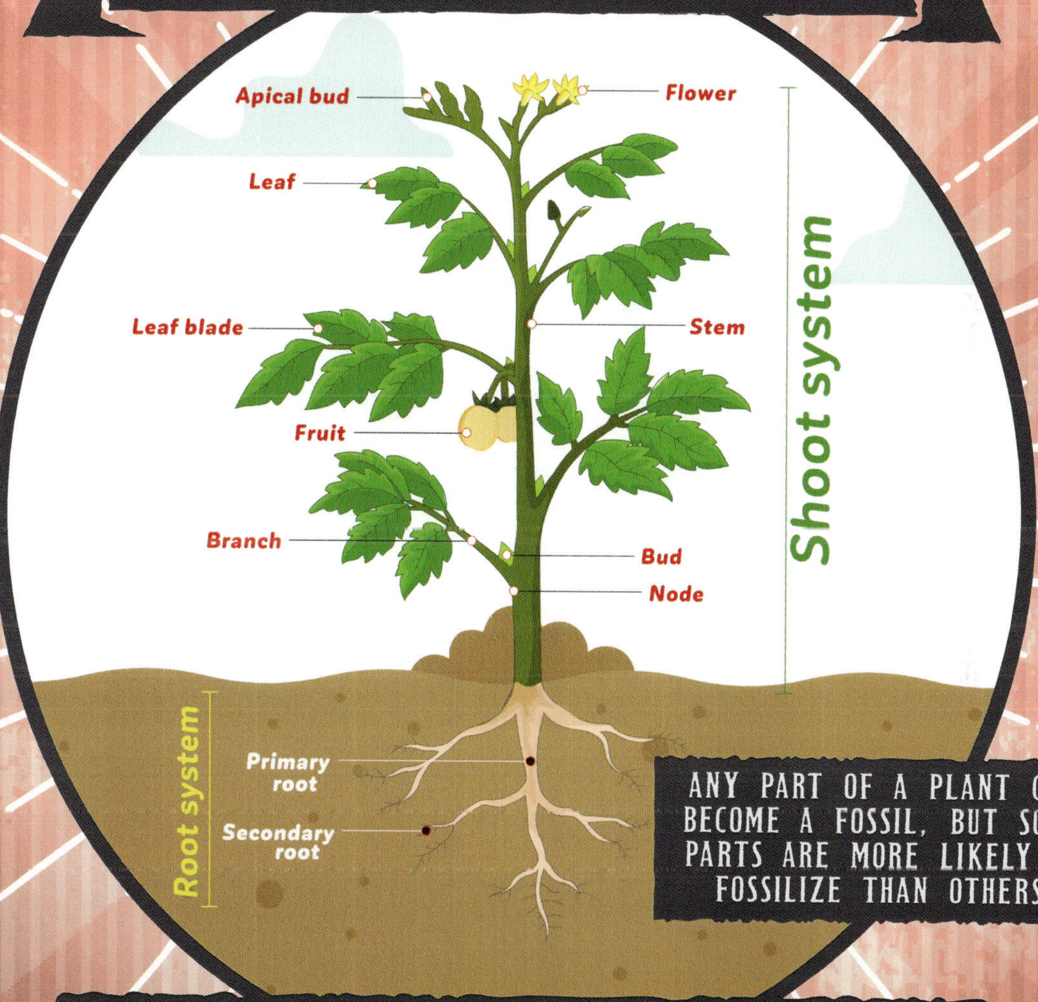

ANY PART OF A PLANT CAN BECOME A FOSSIL, BUT SOME PARTS ARE MORE LIKELY TO FOSSILIZE THAN OTHERS.

Explore More!

FOSSILS ARE NOT COMMON. WHEN MOST PLANTS OR ANIMALS DIE, THEY EITHER BREAK DOWN OR ARE EATEN BY ANOTHER ANIMAL.

THE REMAINS OF PLANTS

There are many kinds of plant fossils. Compressions and impressions happen when plant parts are flattened, or compressed, by the force that changes sediment into rock.

Explore More!

PLANT COMPRESSIONS FORM BEST WHEN THEY ARE FOSSILIZED IN MATTER LIKE CLAY. SEDIMENT FROM CLAY IS VERY FINE AND PRESERVES THE **DETAILS** OF THE PLANT WELL.

THESE LEAF COMPRESSIONS BELONG TO THE *GINKGOITES* GROUP OF PLANTS.

A compression is made of **organic** matter. This means that small bits of the actual plant still remain. Compressions are much more common with plants than animals. Most parts of a plant are flat. Their basic form does not change very much when they are compressed. Since the compression looks much like the original plant, scientists can learn a great deal from these fossils.

THE LEAVES OF THESE FERN PLANTS WERE COMPRESSED MILLIONS OF YEARS AGO.

Impressions are similar to compressions, but they do not contain any living matter. They're just prints of the plant. An impression tells scientists what the outside of the plant looked like.

THESE ANCIENT FERN IMPRESSIONS WERE FOUND IN FRANCE.

Explore More!

DIFFERENT PLANT PARTS ARE MORE LIKELY TO FORM CERTAIN KINDS OF FOSSILS. LEAVES ARE OFTEN PRESERVED AS COMPRESSION FOSSILS. BARK IS MORE LIKELY TO LEAVE AN IMPRESSION FOSSIL.

THE FORM OF THIS PLANT HAS BEEN WELL PRESERVED IN THIS IMPRESSION.

In some cases, a plant makes both a compression and an impression. When a rock layer is split apart, the plant matter sticks to one side of it. This side is the compression, or "part." The other side where the plant left a print of itself in the rock is the impression, or "counterpart."

MORE PLANT FOSSILS

After a plant is buried in sediment, sometimes it breaks down and leaves behind a space in the shape of the plant. This is called a mold. When the mold fills with sediment, a fossil called a cast is formed. Casts and molds can be created by the same plant, like compressions and impressions.

THIS CAST INCLUDES A GROUP OF PLANT STEMS.

Permineralization (puhr-mih-nuh-ruhl-ih-ZAY-shun) can help fossils form. **Minerals** and water drip into a plant buried in sediment. They fill small, empty spaces inside the plant. Sometimes, the minerals take the place of all the plant's parts. The minerals harden. This is called petrification.

THE LIVING PARTS OF THESE PETRIFIED LEAVES WERE REPLACED BY MINERALS AFTER THEY DIED. THE MINERALS THEN TURNED TO STONE.

Explore More!

CASTS AND MOLDS ARE BETTER THAN COMPRESSIONS AND IMPRESSIONS FOR SHOWING WHAT A PLANT LOOKED LIKE, SINCE THE PLANT HAS NOT BEEN FLATTENED.

STUCK IN TIME

Some plants are not fossilized in stone at all. Instead, they are found in matter called amber. Amber comes from fossilized resin—a super sticky matter that is made by trees. If you have ever touched a tree and felt something sticky, that may have been resin.

A DROP OF RESIN RUNS DOWN THE BARK OF A TREE.

Explore More!

SCIENTISTS HAVE FOUND AMBER INCLUSIONS THAT DATE BACK MORE THAN 300 MILLION YEARS!

THIS MOSS HAS BECOME TRAPPED IN AMBER.

When resin drips down the side of the tree, leaves or flowers may get stuck in the resin. Over time, the resin hardens and turns to amber. It is now a fossil called an amber inclusion. Amber does a great job of preserving plants as fossils.

STUDYING PLANT FOSSILS

How do we learn about plant fossils? The study of plant fossils is called paleobotany. "Paleo" means ancient. Botany is the study of plants. The scientists who study plant fossils are paleobotanists. They often study one special kind of plant fossil, such as flowers or trees.

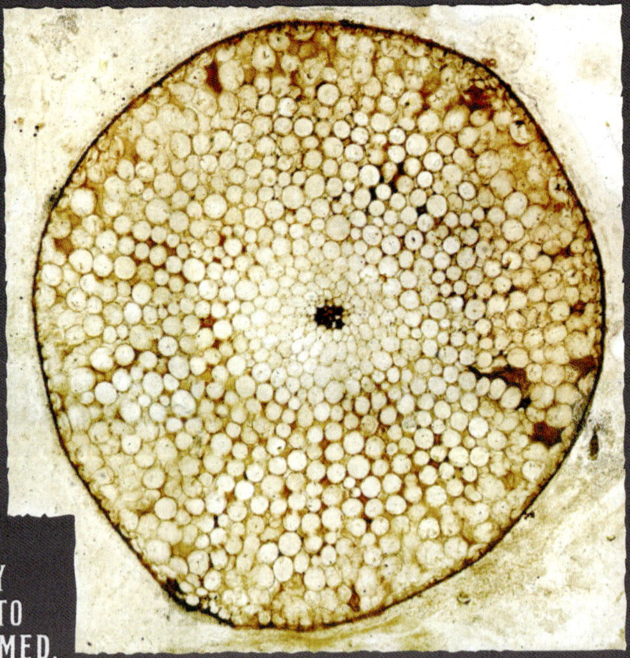

THIS FOSSIL STEM HAS BEEN CUT OPEN. PALEOBOTANISTS MAY STUDY THE SMALLEST PARTS OF FOSSILS TO UNDERSTAND HOW EARLY PLANTS FORMED.

Some paleobotanists study spores and pollen. They are able to study many of these tiny plant parts because they have been well preserved over time. Spore and pollen fossils have helped scientists learn how an area's weather changed over millions of years.

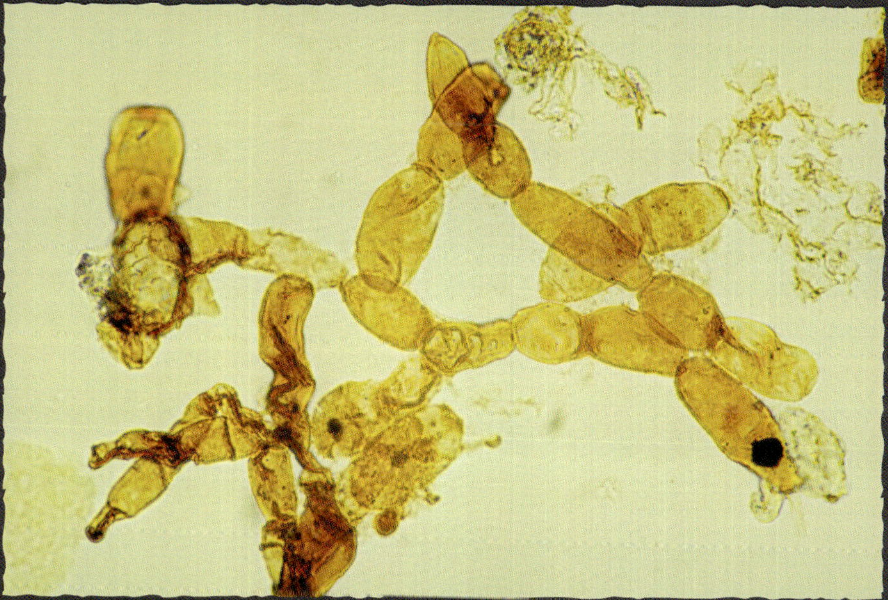

PLANT SPORES LIKE THESE ARE VERY TINY. SCIENTISTS STUDY THEM WITH A MICROSCOPE.

Explore More!

THE STUDY OF PLANT SPORES AND POLLEN IS CALLED PALYNOLOGY.

DIGGING FOR CLUES

Paleobotanists are a bit like **detectives**. They study plant fossils in order to find clues to what Earth was like long ago. It is not easy work. They may start with prospecting, which means searching for the best place to find fossils. The fossils are often very small. If one piece is found, the paleobotanist digs for more.

THIS SCIENTIST HAS JUST FOUND A 67-MILLION-YEAR-OLD TREE LEAF FOSSIL ON A HILLSIDE IN COLORADO.

Digging for fossils can take a long time. Paleobotanists are careful not to break the fossils. Once the fossils are found, they are marked and gently prepared so they can be moved and studied.

THESE FERN FOSSILS WERE FOUND IN SHALE AND ARE WHITE IN COLOR.

Explore More!

ONE OF THE BEST TYPES OF ROCK FOR FINDING PLANT FOSSILS IS SHALE, WHICH COMES FROM MUD AND CLAY.

A FOSSIL'S AGE

Scientists believe that plants have lived on land for more than 440 million years—possibly as many as 500 million years! Each new plant fossil provides clues about how and when plants evolved. They tell us the first forests appeared about 370 million years ago. The first flowering plants, known as angiosperms, may date back more than 149 million years.

Scientists have many ways of figuring out how old a fossil is. Sometimes they look at the rock layer where the fossil was found. This is called superposition. The deeper the rock layer, the older the fossil is.

EARTH'S TIME PERIODS

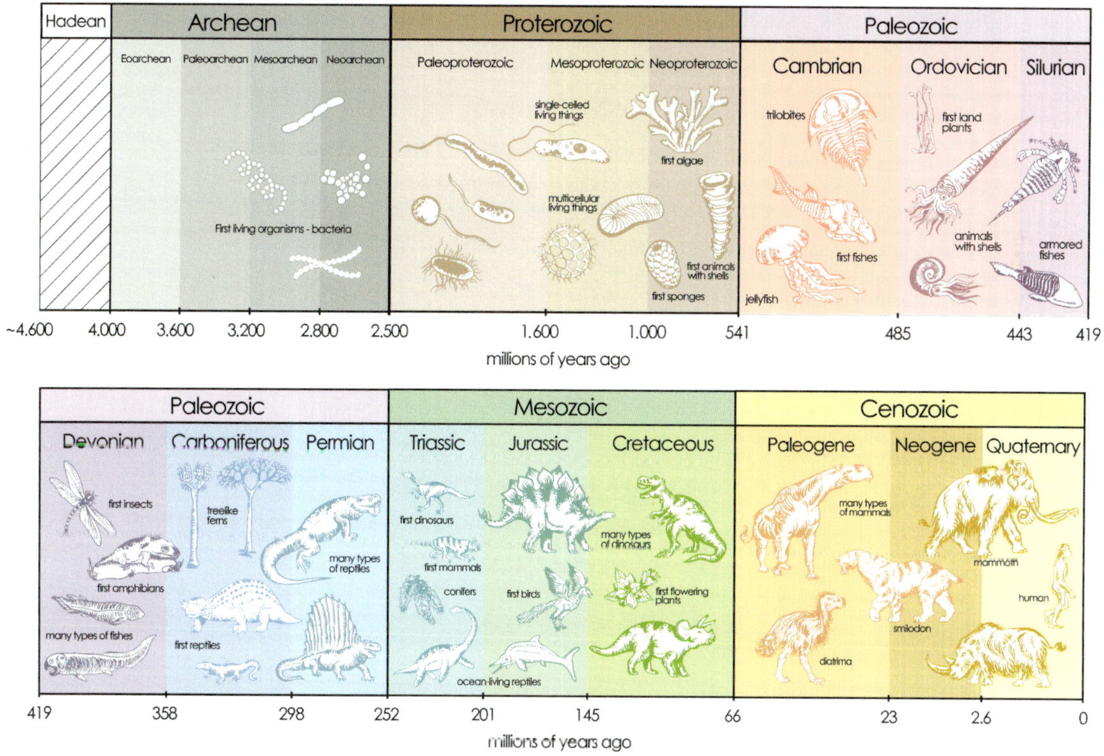

THE EARTH'S HISTORY IS BROKEN DOWN INTO DIFFERENT TIME PERIODS. THE EARLIEST PLANTS PROBABLY CAME ALONG DURING THE ORDOVICIAN PERIOD.

Explore More!

PLANTS STARTED TO GROW ON LAND ONCE THEY HAD A SYSTEM THAT LET THEM MOVE FOOD AND WATER TO ALL THEIR PARTS. THIS IS CALLED A PLANT'S VASCULAR SYSTEM.

FLOWER FOSSILS

All flowering plants, like daisies and sunflowers, are part of the angiosperm group. However, the earliest angiosperm fossils don't show plants with flowers. These plants probably looked more like small bushes or herbs. Paleobotanists have found flower fossils that are 125 million years old. But plants with big, showy flowers didn't start to grow in large numbers until about 95 million years ago.

FLOWER FOSSILS LIKE THESE CAN SHOW SCIENTISTS HOW FLOWERS HAVE CHANGED OVER TIME.

Explore More!

PALEOBOTANISTS ARE STILL TRYING TO LEARN WHETHER ALL FLOWERING PLANTS EVOLVED FROM ONE KIND OF PLANT OR IF THEY COME FROM MANY DIFFERENT PLANTS.

There are not a lot of fossilized flowers. Petals break down too easily to be preserved. To learn what these plants were like, paleobotanists use fossilized fruit, seeds, and pollen.

A GROUP OF FOSSILIZED WATER LILIES. THESE PLANTS EVOLVED WITH LARGE, BRIGHT FLOWERS TO ATTRACT ANIMALS.

ANCIENT ANIMAL FRIENDS

Plants and insects have depended on each other for a long time. Early plants needed help to **pollinate**. They grew flowers to draw animals to them. Insects fed from the flowers. Then the insects flew from flower to flower, pollinating as they went. Plant and insect fossils are often found near each other. By studying them, scientists have learned that plants and insects evolved together.

THESE INSECTS AND LEAVES WERE PRESERVED TOGETHER IN AMBER.

Plant fossils also help scientists learn about the places ancient animals lived. Plant fossils that are found near dinosaur fossils can tell scientists whether the dinosaurs lived in a forest or dry grassland.

FOSSILS CAN TELL US WHAT LIVED ON EARTH AT THE SAME TIME. THIS PALM LEAF WAS FOSSILIZED NEXT TO A SMALL FISH 34 TO 55 MILLION YEARS AGO.

Explore More!

THE OLDEST FOSSIL OF AN INSECT FOUND WITH POLLEN GRAINS IS MORE THAN 100 MILLION YEARS OLD!

PLANTS AND OUR PLANET

Earth's climate—how wet, dry, hot, or cold it is—has changed a lot since it formed more than 4 billion years ago. It's still changing today. Studying plant fossils is one way that we can understand how Earth has changed.

THESE LEAF FOSSILS DATE BACK 245 TO 286 MILLION YEARS AGO. FOSSILS ARE ONE OF THE FEW WAYS SCIENTISTS CAN LEARN WHAT EARTH WAS LIKE SO LONG AGO.

Plants depend on the air and soil to live. They also need different amounts of sun and water. Plant fossils can help scientists figure out what Earth's air, soil, heat, and rain levels were like many years ago. In turn, this could help scientists understand how Earth's climate may change in the future.

Explore More!

SCIENTISTS CAN USE PLANT FOSSILS TO LEARN WHY HUGE NUMBERS OF LIVING THINGS BECAME EXTINCT, OR DIED OUT, AT DIFFERENT TIMES IN HISTORY.

SCIENTISTS CAN FIND LINKS BETWEEN PLANTS AND THE CLIMATE WITH THE HELP OF FOSSILS.

AMAZING ARTIFACTS

If you'd like to check out plant fossils for yourself, there are plenty of places to visit! The National Museum of Natural History in Washington, D.C., opened a new fossil hall in 2019. The huge display of plant, insect, and dinosaur fossils shows how important it is to understand our past.

SCIENTISTS LIKE RALF WERNEBURG CONTINUE TO MAKE DISCOVERIES LIKE THIS 298-MILLION-YEAR-OLD FOSSIL OF A HORSETAIL PLANT.

Explore More!

TAKE A HIKE AROUND PETRIFIED FOREST NATIONAL PARK IN ARIZONA TO SEE ONE OF THE WORLD'S LARGEST PETRIFIED FORESTS, AS WELL AS OVER 200 OTHER TYPES OF PLANT FOSSILS.

At the Florida Museum of Natural History in Gainesville, Florida, you can view plant fossils with histories that go back 425 million years! Each fossil tells a part of the story of how life on Earth has changed over millions of years.

GLOSSARY

algae Living things that grow in water and make their own food, much like plants do.

detail A small part.

detective Someone whose job is to find out information about someone or something.

evolve To change slowly over a long period of time.

mineral Matter found in nature that is not living.

organic Having to do with living things.

pollen A fine yellow dust made by plants.

pollinate To carry pollen to a plant.

preserve To keep safe.

remains Parts of a living thing that are left behind after it dies.

sediment Matter, such as stones and sand, that is carried onto land or into the water by wind, water, or land movement.

spore A small body made by a plant that can grow into another plant.

For More Information

BOOKS

Alkire, Jessie. *Discovering Fossils*. Minneapolis, MN: Abdo Publishing, 2019.

Barnham, Kay. *The Amazing Life Cycle of Plants*. Hauppauge, NY: Barron's, 2018.

Martin, Claudia. *Fossils*. London, England: QED Publishing, 2018.

WEBSITES

American Museum of Natural History: Paleontology
www.amnh.org/explore/ology/paleontology
Find out what it's like to dig for fossils through videos, games, and activities.

DK Find Out: Parts of a Flower
www.dkfindout.com/us/animals-and-nature/plants/parts-flower/
Explore the different parts of a flower and what their jobs are.

Ducksters: Fossils
www.ducksters.com/science/earth_science/fossils.php
Discover more fun facts about fossils, then check your knowledge with a quiz.

INDEX